Susie Barstow Skelding

Flowers from dell and bower

Susie Barstow Skelding

Flowers from dell and bower

ISBN/EAN: 9783337113490

Printed in Europe, USA, Canada, Australia, Japan

Cover: Foto ©berggeist007 / pixelio.de

More available books at **www.hansebooks.com**

FLOWERS
FROM
DELL AND BOWER

POEMS ILLUSTRATED

BY

SUSIE BARSTOW SKELDING

*Author of "Flowers from Hill and Dale," "The Flower-Songs Series,"
etc., etc.*

NEW YORK
WHITE, STOKES, & ALLEN
1886

ILLUSTRATIONS.

BY SUSIE BARSTOW SKELDING.

	PAGE
JACQUEMINOT ROSES, . .	*Frontispiece*
TRAILING ARBUTUS, ...	. 23
WHITE DAISIES AND GRASSES, . .	33
WILD ROSES, ...	. 43
EASTER LILIES, ...	. 53

(*With fac-simile of manuscript, by Lucy Larcom.*)

SWEET-PEAS, ...	. 63
VIOLETS, . .	. 73
MOSS ROSES, ...	. 83
JONQUILS AND CROCUSES, .	93
PINK AND WHITE AZALEAS, . .	103
WHITE LILIES, ...	. 113
PALE YELLOW ROSES, ...	. 123

The editor acknowledges the courtesy of Messrs. Roberts Brothers and D. Appleton & Co. for the use of poems by prominent writers.

CONTENTS.

	PAGE
A GARDEN IDYL,	15
(*Frederick Locker.*)	
QUOTATION,	17
(*Campbell.*)	
THE ROSE'S PRIDE,	18
(*Sir Richard Fanshawe.*)	
INCENSE OF FLOWERS,	19
(*Robert Leighton.*)	
THE TRUE AND THE FALSE,	20
(*Shakespeare.*)	
THE ARBUTUS,	25
(*Fac-simile of manuscript, by Helen Jackson—"H. H."*)	
ON OBSERVING A BLOSSOM ON THE FIRST OF FEBRUARY,	27
(*Coleridge.*)	
SPRING IN THE LAP OF WINTER,	28
(*Anonymous.*)	
LATE SPRING,	29
(*Southey.*)	
TO A FLOWER,	30
(*Anonymous.*)	
TO A MOUNTAIN DAISY,	35
(*Burns.*)	
THE DAISIE,	38
(*Chaucer.*)	
MOSSGIEL,	39
(*Wordsworth.*)	

CONTENTS.

	PAGE
FROM EASTER MESSENGERS,	40
(*Lucy Larcom.*)	
THE WILD ROSES,	45
(*Elizabeth D. Bullock.*)	
SONG,	48
(*Arthur O'Shaughnessy.*)	
THE ROSE,	49
(*Lovelace.*)	
QUOTATION,	49
(*Thomson.*)	
FROM EASTER BELLS,	55
(*Fac-simile of manuscript, by Helen Jackson—"H. H."*)	
ANGELS ROLL THE ROCK AWAY,	56
(*The Rev. Thomas Scott.*)	
WELCOME, O DAY!	57
(*William Allen, D.D.*)	
QUOTATION,	58
(*Wordsworth.*)	
QUOTATION,	58
(*Tennyson.*)	
QUOTATION,	58
(*Cowley.*)	
THE LILY,	59
(*J. G. Percival.*)	
SWEET-PEAS,	65
(*Keats.*)	
FROM THE FLOWER,	65
(*George Herbert.*)	
THE GARDEN,	66
(*Andrew Marvell.*)	
FROM EASTER MESSENGERS,	70
(*Lucy Larcom.*)	

CONTENTS.

	PAGE
SONG,	75
(William Cullen Bryant.)	
VIOLETS,	77
(Leigh Hunt.)	
THE CLOSE OF SPRING,	78
(Charlotte Smith.)	
FLOWERS,	79
(Barry Cornwall.)	
THE VIOLET,	80
(Translated from Goethe.)	
SONNET,	85
(Spenser.)	
"DEEP IN THE SNOW'S BED BURY THE ROSE,"	86
(Swinburne.)	
"SEE, SEE THE FLOWERS THAT BELOW,"	89
(Giles Fletcher.)	
"THE ROSES FEARFULLY ON THORNS DID STAND,"	89
(Shakespeare.)	
THE FUNERAL RITES OF THE ROSE,	90
(Herrick.)	
MAY-TIDE,	95
(Lewis Morris.)	
THE CROCUS,	97
(Mary Howitt.)	
FROM EASTER MESSENGERS,	97
(Lucy Larcom.)	
TO A CROCUS,	98
(Bernard Barton.)	
SONG IN PRAISE OF SPRING,	105
(Barry Cornwall.)	
FLOWERS,	107
(Anonymous.)	

CONTENTS.

	PAGE
FROM EASTER MESSENGERS,	115
(*Fac-simile of manuscript, by Lucy Larcom.*)	
A DIALOGUE FROM SOUL GARDENING,	116
(*Dora Greenwell.*)	
LILIES,	117
(*Leigh Hunt.*)	
THE LILY,	118
(*Coleridge.*)	
I SEND THE LILIES GIVEN TO ME,	119
(*Byron.*)	
STANZA,	125
(*Quarles.*)	
"O HAVE YOU SEEN?"	126
(*L. E. Landon.*)	
"THE ROSE SAID,"	126
(*Augusta Webster.*)	
THE FALLING ROSE,	127
(*William Cox Bennett.*)	
"WOO ON, WITH ODOR WOOING ME,"	128
(*George MacDonald.*)	
"LIVE ALL THY SWEET LIFE THROUGH,"	128
(*Christina G. Rossetti.*)	

JACQUEMINOT ROSES.

A GARDEN IDYLL.

There are plenty of roses (the patriarch speaks),
But alas not for me, on your lips and your cheeks;
Sweet Maiden, rose-laden—enough and to spare—
Spare, O spare me the rose that you wear in your hair.

We have loiter'd and laugh'd in the flowery croft,
    We have met under wintry skies;
Her voice is the dearest voice, and soft
    Is the light in her wistful eyes;
It is sweet in the silent woods, among
    Gay crowds, or in any place
To hear her voice, to gaze on her young
    Confiding face.

Forever may roses divinely blow,
    And wine-dark pansies charm
By the prim box-path where I felt the glow
    Of her dimpled, trusting arm,
And the sweep of her silk as she turn'd and smiled
    A smile as fair as her pearls;
The breeze was in love with the darling child,
    As it moved her curls.

She showed me her ferns and woodbine sprays,
    Fox-glove and jasmine stars,
A mist of blue in the beds, a blaze
    Of red in the celadon jars:
And velvety bees in convolvulus bells,
    And roses of beautiful June—
Oh, who would think the Summer spells
    Could die so soon!

For a glad song came from the milking shed,
    On a wind of that Summer south,
And the green was golden above her head,
    And a sunbeam kissed her mouth;
Sweet were the lips where that sunbeam dwelt—
    And the wings of Time were fleet
As I gazed; and neither spoke, for we felt
    Life was so sweet!

And the odorous limes were dim above
    As we leant on a drooping bough;
And the darkling air was a breath of love,
    And a witching thrush sang " Now!"
For the sun dropt low, and the twilight grew
    As we listen'd, and sigh'd, and leant—
That day was the sweetest day—and we knew
    What the sweetness meant.
Frederick Locker.

WHEN Love came first to earth, the Spring
Spread rose-beds to receive him.
Campbell.

THE ROSE'S PRIDE.

Thou blushing rose, within whose virgin leaves
  The wanton wind to sport himself presumes,
Whilst from their rifled wardrobe he receives
  For his wings purple, for his breath perfumes!

Blown in the morning, thou shalt fade ere noon;
  What boots a life which in such haste forsakes thee?
Thou'rt wondrous frolic, being to die so soon,
  And passing proud a little color makes thee.
                              Sir Richard Fanshawe.

INCENSE OF FLOWERS.

This rich abundance of the rose, its breath,
  On which I almost think my soul could live,
This sweet ambrosia, which even in death
      Its leaves hold on to give—

Whence is it?   From dark earth or scentless air?
  Or from the inner sanctuaries of heaven?
We probe the branch, the root, no incense there—
      O God, whence is it given?

Is it the essence of the morning dew,
  Or distillation of a purer sphere—
The breath of the immortals coming through
      To us immortals here?

Exquisite mystery, my heart devours
  The living inspiration, and I know
Sweet revelations with the breath of flowers
      Into our beings flow.

Robert Leighton.

THE TRUE AND THE FALSE.

O, how much more doth beauty beauteous seem
    By that sweet ornament which truth doth give!
The rose looks fair, but fairer we it deem
    For that sweet odor which doth in it live.

The canker blooms have full as deep a dye
    As the perfumed tincture of the roses,
Hang on such thorns, and play as wantonly
    When Summer's breath their maskèd buds discloses.

But, for their virtue only is their show,
    They live unwoo'd and unrespected fade,
Die to themselves. Sweet roses do not so:
    Of their sweet deaths are sweetest odors made;

And so of you, beauteous and lovely youth,
    When that shall fade, thy verse distils your truth.
                      Shakespeare.

TRAILING ARBUTUS.

The Arbutus.

Of all the Spring's beloved,
  Oh fair and fickle Spring,
Not one except Arbutus,
  Can trust what she will bring.
The rest, one year, get blossoms,
  In punctual sunshine bright
The next, they wait and shiver
  And droop in icy blight.
Arbutus, little, and lover
  Has forged for Spring such chain,
However far she wanders,
  She's back on time again.
And whether snows have melted,
  Or lie all solid white,
Arbutus blossoms blossom,
  Their rosy cups all right.
And happy youths and maidens
  Its secret haunts who know,
Go confident to seek it,
  And find it 'neath the snow.
Of all the Spring's beloved
  Of all that she can bring,
If she'll give us Arbutus,
  The rest may wait all Spring!

H. H.

ON OBSERVING A BLOSSOM ON THE FIRST OF FEBRUARY.

Sweet Flower! that peeping from thy russet stem
Unfoldest timidly (for in strange sort
This dark, frieze-coated, hoarse, teeth-chattering Month
Hath borrowed Zephyr's voice, and gazed on thee
With blue, voluptuous eye), alas, poor Flower!
These are but flatteries of the faithless year.
Perchance, escaped its unknown polar cave,
E'en now, the keen North-East is on its way.
Flower that must perish! Shall I liken thee
To some sweet girl, of too, too rapid growth,
Nipped by consumption 'mid untimely charms,
Or to Bristowa's bard, the wondrous boy!
An amaranth, which earth scarce seemed to own,
Till disappointment came, and pelting wrong
Beat it to Earth? or with indignant grief
Shall I compare thee to poor Poland's hope,
Bright flower of hope, killed in the opening bud?
Farewell, sweet blossom! better fate be thine.
Coleridge.

SPRING IN THE LAP OF WINTER.

The mist still hovers round the distant hills;
But the blue sky above us has a clear
And pearly softness; not a white speck lies
Upon its breast; it is a crystal dome.
There is a quiet charm about this morn
Which sinks into the soul. No gorgeous colors
Has the undraperied earth, but yet she shows
A vestal brightness: not the voice is heard
Of sylvan melody, whether of birds
Intent on song, or bees mingling their music
With their keen labor; but the twittering voice
Of chaffinch, and the wild unfrequent note
Of the lone woodlark, and the minstrelsy
Of the blest robin, have a potent spell
Chirping away the silence; not the perfume
Of violets scents the gale, nor apple-blossom,
Nor satiating bean-flower; the fresh breeze
Itself is purest fragrance. Light and air
Are ministers of gladness; where these spread,
Beauty abides, and joy: where'er Life is
There is no melancholy.
                              Anonymous.

LATE SPRING.

Thou lingerest, Spring, still wintry is the scene;
  The fields their dead and sapless russet wear;
  Scarce does the glossy Celandine appear
Starring the sunny bank, or, early green,
The Elder yet its circling tufts put forth;
  The sparrow tenants still the eave-built nest,
  Where we should see our martin's snowy breast
Oft darting out. The blasts from the bleak North
  And from the keener East still frequent blow;
  Sweet Spring, thou lingerest, and it should be so—
Late let the fields and gardens blossom out!
  Like man, when most with smiles thy face is drest,
  'Tis to deceive, and he who knows you best,
When most ye promise, even most will doubt.

Southey.

TO A FLOWER.

Dawn, gentle flower,
  From the morning earth!
We will gaze and wonder
  At thy wondrous birth!

Bloom, gentle flower!
  Lover of the light,
Sought by wind and shower,
  Fondled by the night!

Fade, gentle flower!
  All thy white leaves close;
Having shown thy beauty,
  Time 'tis for repose.

Die, gentle flower,
  In the silent sun!
So,—all pangs are over,
  All thy tasks are done.

Day hath no more glory,
  Though he soars so high;
Thine is all man's story,
  Live,—and love,—and die!

Anonymous.

WHITE DAISIES AND GRASSES.

TO A MOUNTAIN DAISY.

ON TURNING ONE DOWN WITH A PLOW.

Wee, modest, crimson-tipped flower,
Thou's met me in an evil hour,
For I maun crush amang the stoure
    Thy slender stem;
To spare thee now is past my power,
    Thou bonnie gem.

Alas! it's no thy neebor sweet,
The bonnie lark, companion meet,
Bending thee 'mang the dewy weet,
    Wi' speckled breast,
When upward springing, blithe to greet
    The purpling east.

TO A MOUNTAIN DAISY.

Cauld blew the bitter, biting north
Upon thy early, humble birth;
Yet cheerfully thou glinted forth
    Amid the storm,
Scarce reared above the parent earth
    Thy tender form.

The flaunting flowers our gardens yield,
High sheltering woods and wa's maun shield,
But thou, beneath the random bield
    O' clod or stane
Adorns the histie stibble-field,
    Unseen, alane.

There, in thy scanty mantle clad,
Thy snawie bosom sunward spread,
Thou lifts thy unassuming head
    In humble guise;
But now the share uptears thy bed,
    And low thou lies!

Such fate to suffering worth is given,
Who lang with wants and woes has striven,
By human pride or cunning driven
    To misery's brink,
Till, wrenched of every stay but Heaven,
    He ruined, sink!

TO A MOUNTAIN DAISY.

Even thou, who mourn'st the daisy's fate
That fate is thine—no distant date,
Stern Ruin's plowshare drives elate
    Full on thy bloom,
Till, crushed beneath the furrow's weight,
    Shall be thy doom!

Burns.

THE DAISIE.

DAISIE of light! very ground of comfort!
The sunnis doughtir ye hight, as I rede,
For when he westrith, farwell your disport;
By your nature anone, right for pure drede
Of the rude Night, that with his boistous wede
Of derkenesse shadowith our hemisphere,
Then closin ye, my liv'is ladie dere.

Daunying the daie unto his kind resort,
And Phœbus your fethir with his stremes rede
Adorneth the morrowe, consuming the sort
Of mistie cloudes, that wouldin ovirlede
True humble hertis with ther mistie hede,
Nere comfort adaies, when your eyin clere
Disclose and sprede, my liv'is ladie dere.

Je vouldray; but the grete God disposeth
And makith casueil by His providence
Soche thing as mannis frele wit purposeth,
All for the best, if that your conscience
Not grutche it, but in humble pacience
It receve; for God saith withoutin fable,
A faithfull herte evir is acceptable.
Chaucer.

MOSSGIEL.

"THERE," said a stripling, pointing with much pride,
    Towards a low roof, with green trees half-concealed,
"Is Mossgiel farm; and that's the very field
Where Burns plough'd up the daisy!" Far and wide
A plain below stretch'd seaward; while, descried
    Above sea clouds, the peaks of Arran rose;
    And, by that simple notice, the repose
Of earth, sky, sea and air was vivified.
    Beneath the random field of clod or stone
Myriads of daisies here shone forth in flower
Near the lark's nest, and in their natural hour
    Have passed away; less happy than the one
That, by the unwilling ploughshare, died to prove
The tender charm of poetry and love.
                              Wordsworth.

From EASTER MESSENGERS.

A MESSAGE from the Grasses
   Is whispered by the breeze :—
" O give thyself to service meek,
   And lowly ministries ;
Contented, though unnoticed,
   To weave a tender grace
Around the toiler's trodden path—
   The poor man's dwelling-place."

And softly sighs the Daisy—
   " Although thy lot be low,
The humblest wild-flower of the field
   In light from heaven must grow.
The ploughboy, if he watches,
   May see the King go by :
At work, at rest, still keep thy soul
   Wide open to the sky !"

Lucy Larcom.

WILD ROSES.

THE WILD ROSES.

I WALKED in the joyous morning,
    The morning of June and life,
Ere the birds had ceased to warble
    Their sweetest of love and strife.

I walked alone in the morning,
    And who so glad as I,
When I saw the pale wild roses
    Hang from the branch on high?

Fairer than stars were the roses,
    Faint was the fragrance and rare,
Not any flower in the garden
    Could with those roses compare.

But the day was all before me,
  The tumult of youth's delight;
Why bear a burden of roses
  Before the calm of the night?

Let them stay awhile to gladden
  The air and the earth below,
With tender beauty and sweetness
  They cannot choose but bestow.

So I kissed the roses, and lightly
  I breathed of their breath divine;
It is time when I come back, I said,
  To make the sweet roses mine.

I went in the gladsome morning,—
  I said we part for an hour;
The branch of wild roses trembled,
  The dew was on every flower.

I returned in the joyless evening,
  I yearned with passion then,
For the pale and peerless roses
  I never should see again.

For another had taken delight
  In color and perfume rare,
And another hand had gathered
  My roses beyond compare.

I may wander East, may wander West,
  Wherever the sun doth shine,
I never shall find the wild roses,
  The roses I thought were mine.
Elizabeth D. Bullock.

SONG.

When the rose came, I loved the rose,
　　And thought of none beside,
Forgetting all the other flowers,
　　And all the others died;
And morn, and noon, and sun, and showers,
　　And all things loved the rose,
Who only half returned my love,
　　Blooming alike for those.

I was the rival of a score
　　Of loves on gaudy wing,
The nightingale I would implore
　　For pity not to sing;
Each called her his, still I was glad
　　To wait or take my part;
I loved the rose—who might have had
　　The fairest lily's heart.
　　　　　　　　Arthur O'Shaughnessy.

THE ROSE.

Sweet, serene, sky-like flower,
Haste to adorn her bower;
From thy long cloudy bed
Shoot forth thy damask head.

Vermillion ball that's given
From lip to lip in heaven;
Love's couch's coverlid;
Haste, haste to make her bed.

See, rosy is her bower,
Her floor is all thy flower;
Her bed a rosy nest,
A bed of roses prest.
Lovelace.

Flowers of all hue, their queen the bashful rose.
Thomson.

EASTER LILIES.

(51)

Gone feud and fray,
  Faded away.
Forgot today, as glad we kneel,
  With prayers that heal;
    The earth is old;
      Clear eyed and bold,
Faith sees her way, and keeps her hold;
Christ risen, shines in darkest skies;
    Ring Easter bells!
      Loud brazen bells,
  Sweet silver bells of centuries!

                    Helen Jackson.
                      (H.H.)

ANGELS ROLL THE ROCK AWAY!

ANGELS roll the rock away!
Death, yield up the mighty prey!
See! the Saviour quits the tomb,
Glowing with immortal bloom.
   Hallelujah! hallelujah!
Christ the Lord is risen to-day.

Shout, ye seraphs! angels, raise
Your eternal song of praise!
Let the earth's remotest bound
Echo to the blissful sound!
   Hallelujah! hallelujah!
Christ the Lord is risen to-day.

Holy Father, holy Son,
Holy Spirit, three in One,
Glory as of old to thee,
Now and evermore shall be:
   Hallelujah! hallelujah!
Christ the Lord is risen to-day.

—The Rev. Thomas Scott.

WELCOME, O DAY!

Welcome, O day! in dazzling glory bright!
Emblem of yet another day most blest,
When all Christ's friends with him in heaven shall rest;
   For on this day, in his recovered might,
   The sleeper waked to see this morning's light,—
"The Son of God!" glad angel hosts attest:
So, when alive, most fully shown confest;
   For on this day he took his heavenward flight.
When, therefore, our glad eyes this morning's sun
   See rising on the earth, we'll lift our thought
   To him who by his death our life hath bought,
And, Victor, King, for us a crown hath won.
   It e'er shall be a day of sweetest joy,
   Till we shall see our Lord in yonder sky!

                              —*William Allen, D.D.*

A LILY flower,
The old Egyptian's emblematic mark
Of joy immortal and of pure affection.
Wordsworth.

Now folds the lily all her sweetness up,
And slips into the bosom of the lake;
So fold thyself, my dearest, thou, and slip
Into my bosom, and be lost in me.
Tennyson.

THE virgin lilies in their white,
Clad but with the lawn of almost naked white.
Cowley.

THE LILY.

I HAD found out a sweet green spot,
   Where a Lily was blooming fair;
The din of the city disturbed it not,
But the spirit, that shades the quiet cot
   With its wings of Love, was there.

I found that Lily's bloom
   When the day was dark and chill:
It smiled, like a star in the misty gloom,
And it sent abroad a soft perfume,
   Which is floating around me still.

I sat by the Lily's bell,
   And watched it many a day:—
The leaves, that rose in a flowing swell,
Grew faint and dim, then drooped and fell,
   And the flower had flown away.

I looked where the leaves were laid
   In withering paleness, by,
And, as gloomy thoughts stole on me, said,
There is many a sweet and blooming maid,
   Who will soon as dimly die.

J. G. Percival.

SWEET-PEAS.

SWEET-PEAS.

Here are sweet-peas, on tiptoe for a flight;
With wings of gentle flush o'er delicate white,
And taper fingers catching at all things,
To bind them all about with tiny rings.

Keats.

From THE FLOWER.

How fresh, O Lord, how sweet and clean
   Are thy returns, e'en as the flowers in Spring;
To which, besides their own demean,
   The late-past frosts tributes of pleasure bring.
     Grief melts away
     Like snow in May;
As if there were no such cold thing.

George Herbert.

THE GARDEN.

How vainly men themselves amaze,
To win the palm, the oak, or bays;
And their incessant labors see
Crown'd from some single herb, or tree,
Whose short and narrow verged shade
Does prudently their toils upbraid;
While all the flowers and trees do close,
To weave the garlands of Repose.

Fair Quiet, have I found thee here,
And Innocence, thy sister dear!
Mistaken long, I sought you then
In busy companies of men.
Your sacred plants, if here below,
Only among the plants will grow.
Society is all but rude
To this delicious solitude.

THE GARDEN.

No white nor red was ever seen
So am'rous as this lovely green.
Fond lovers, cruel as their flame,
Cut in these trees their mistress' name.
Little, alas, they know or heed
How far these beauties her exceed!
Fair trees, where'er your barks I wound,
No name shall but your own be found.

When we have run our passion's heat,
Love hither makes his best retreat.
The gods, who mortal beauty chase,
Still in a tree did end their race.
Apollo hunted Daphne so,
Only that she might laurel grow.
And Pan did after Syrinx speed,
Not as a nymph, but for a reed.

What wondrous life is this I lead!
Ripe apples drop about my head;
The luscious clusters of the vine
Upon my mouth do crush their wine
The nectarine, the curious peach,
Into my hands themselves do reach.
Stumbling on melons, as I pass,
Ensnar'd with flowers, I fall on grass.

Meanwhile the mind, from pleasure less,
Withdraws into its happiness;
The mind, that ocean where each kind
Does straight its own resemblance find;
Yet it creates, transcending these,
Far other worlds, and other seas;
Annihilating all that's made
To a green thought in a green shade.

Here, at the fountain's sliding foot,
Or at some fruit-tree's mossy root,
Casting the body's vest aside,
My soul into the boughs does glide:
Here, like a bird, it sits and sings,
Then whets, and claps its silver wings;
And, till prepared for longer flight,
Waves in its plumes the various light.

Such was that happy garden-state,
While man there walk'd without a mate:
After a place so pure and sweet,
What other help could yet be meet!
But 'twas beyond a mortal's share
To wander solitary there:
Two Paradises are in one,
To live in Paradise alone.

THE GARDEN.

How well the skillful gard'ner drew
Of flowers and herbs, this dial new:
Where, from above, the milder sun
Does through a fragrant zodiac run;
And, as it works, th' industrious bee
Computes his time as well as we.
How could such sweet and wholesome hours
Be reckon'd but with herbs and flowers?
Andrew Marvell.

From EASTER MESSENGERS.

THE Sweet-pea flutters upward—
   Almost a flower with wings—
Her tendrils steadying her flight;
   Ascending while she clings.
Through fragrant mists she murmurs
   "O make life sweet with prayer!
Hold firmly by the human ties,
   But breathe in heavenly air!"
Lucy Larcom.

VIOLETS.

SONG.

Dost thou idly ask to hear
    At what gentle seasons
Nymphs relent, when lovers near
    Press the tenderest reasons?
Ah, they give their faith too oft
    To the careless wooer;
Maidens' hearts are always soft:
    Would that men's were truer!

Woo the fair one when around
    Early birds are singing;
When o'er all the fragrant ground,
    Early herbs are springing:
When the brookside, bank, and grove,
    All with blossoms laden,
Shine thy beauty, breathe of love,—
    Woo the timid maiden.

Woo her when, with rosy blush,
    Summer eve is sinking;
When on rills that softly gush,
    Stars are softly winking;

SONG.

When through boughs that knit the bower
   Moonlight gleams are stealing;
Woo her, till the gentle hour
   Wake a gentler feeling.

Woo her, when autumnal dyes
   Tinge the woody mountain;
When the dropping foliage lies
   In the weedy fountain;
Let the scene, that tells how fast
   Youth is passing over,
Warn her, ere her bloom is past,
   To secure her lover.

Woo her when the north winds call
   At the lattice nightly;
When, within the cheerful hall,
   Blaze the fagots brightly;
While the wintry tempest round
   Sweeps the landscape hoary,
Sweeter in her ear shall sound
   Love's delightful story.

—William Cullen Bryant.

VIOLETS.

We are violets blue,
   For our sweetness found
Careless in the mossy shades,
   Looking on the ground.
Love's dropp'd eyelids and a kiss,—
Such our breath and blueness is.

Io, the mild shape
   Hidden by Jove's fears,
Found us first i' the sward, when she
   For hunger stoop'd in tears.
"Wheresoe'er her lips she sets,"
Jove said, "be breaths call'd Violets."
Leigh Hunt.

THE CLOSE OF SPRING.

The garlands fade that Spring so lately wove,
    Each simple flower which she had nursed in dew,
Anemones that spangled every grove—
    The primrose wan, and harebell mildly blue.
No more shall violets linger in the dell,
    Or purple orchis variegate the plain,
Till Spring again shall call forth every bell,
    And dress with humid hands her wreaths again—
Ah, poor humanity! so frail, so fair,
    Are the fond visions of thine early day,
Till tyrant passion and corrosive care
    Bid all thy fairy colors fade away!
Another May new buds and flowers shall bring;
Ah! why has happiness no second Spring?
Charlotte Smith.

FLOWERS.

WE have left behind us
The riches of the meadows,—and now come
To visit the virgin primrose where she dwells,
'Midst harebells and the wild-wood hyacinths.
'Tis here she keeps her court. Dost see yon bank
The sun is kissing? Near,—go near! for there,
('Neath those broad leaves, amidst yon straggling
    grasses)
Immaculate odors from the violet
Spring up for ever! Like sweet thoughts that come
Winged from the maiden fancy, and fly off
In music to the skies, and there are lost,
These ever-steaming odors seek the sun,
And fade in the light he scatters.
                            Barry Cornwall.

THE VIOLET.

A Violet blossom'd on the green,
With lowly stem, and bloom unseen;
    It was a sweet, low flower.
A shepherd maiden came that way,
With lightsome step and aspect gay,
    Came near, came near,
Came o'er the green with song.

Ah! thought the Violet, might I be
The fairest flower on all the lea,
    Ah! but for one brief hour:
And might be plucked by that dear maid,
And gently on her bosom laid,
    Ah! but, ah! but
A few dear moments long.

Alas! the maiden, as she pass'd,
No eye upon the Violet cast;
    She crush'd the poor wee flower;
It sank, and, dying, heaved no sigh,
And if I die, at least I die
    By her, by her,
Beneath her feet I die.
               Translated from Goethe.

MOSS ROSES.

SONNET.

Sweet is the Rose, but growes upon a brere;
Sweet is the Juniper, but sharpe his bough;
Sweet is the Eglantine, but pricketh nere;
Sweet is the Firbloom, but his branch is rough;
Sweet is the Cypress, but his rind is tough,
Sweet is the Nut, but bitter is his pill;
Sweet is the Broome-flowere, but yet sowre enough;
And sweet is Moly, but his roote is ill.
So every sweet with sowre is tempered still,
That maketh it to be coveted the more:
For easie things that may be got at will,
Most sorts of men doe set but little store.
Why then should I account of little pain,
That endless pleasure shall unto me gaine?

Spenser.

DEEP in the snow's bed bury the rose.
  Bury her deeper
  Than any sleeper;
  Sweet dreams will keep her
   All day, all night;
  Though sleep benumb her,
  And time overcome her,
  She dreams of Summer,
   And takes delight,
  Dreaming and sleeping
  In love's good keeping,
  While rain is weeping
   And no leaves cling;
  Winds will come bringing her
  Comfort, and singing her
Stories, and songs, and good news of the Spring.

DEEP IN THE SNOW'S BED.

    Draw the white curtain
    Close, and be certain
    She takes no hurt in
       Her soft, low bed:
    She feels no colder,
    And grows no older,
    Though snows enfold her
       From foot to head;
    She turns not chilly
    Like weed and lily
    In marsh or hilly
       High watershed,
    Or green soft island,
    In lake of highland;
She sleeps awhile, and she is not dead.

    For all the hours,
    Come sun, come showers,
    Are friends of flowers,
       And fairies all;
    When frost entrapped her
    They came and lapped her
    In leaves, and wrapped her
       With shroud and pall;
    In red leaves wound her,
    With dead leaves bound her
    Dead brows, and round her

A death-knell rang;
Rang the death-bell for her,
Sang, "Is it well for her,
Well, is it well with you, rose?" they sang.

O what and where is
The rose now, fairies,
So shrill the air is,
 So wild the sky?
Poor last of roses,
Her worst of woes is
The noise she knows is
 The Winter's cry;
His hunting hollo
Has scared the swallow;
Fain would she follow
 And fain would fly;
But wind unsettles
Her poor last petals;
Had she but wings, and she would not die.
Swinburne.

SEE, see the flowers that below
Now as fresh as morning blow,
And of all the virgin rose,
That as bright Aurora shows:
How they all unleavèd lie,
Losing their virginity;
Like unto a Summer shade,
But now born and now they fade.
Everything doth pass away;
There is danger in delay;
Come, come, gather then the rose;
Gather it, or it you lose.
Giles Fletcher.

THE roses fearfully on thorns did stand,
One blushing shame, another white despair,
A third, nor red nor white, had stolen of both,
And to his robbery had annexed thy breath.
Shakespeare.

THE FUNERAL RITES OF THE ROSE.

THE rose was sick, and, smiling, died;
And, being to be sanctified,
About the bed there sighing stood
The sweet and flowery sisterhood.
Some hung the head, while some did bring,
To wash her, water from the spring;
Some laid her forth, while others wept,
But all a solemn fast there kept.
The holy sisetrs, some among
The sacred dirge and trental sung:
But ah! what sweets smelt everywhere,
As heaven had spent all perfumes there!
At last, when prayers for the dead,
And rites, were all accomplishéd,
They, weeping, spread a lawny loom
And closed her up as in a tomb.
Herrick.

JONQUILS AND CROCUSES.

MAY-TIDE.

This is the hour, the day,
  The time, the season sweet!
  Quick, hasten, laggard feet!
Brook not delay!
Love flies, youth passes, May-tide will not last:
Forth, forth, while yet 'tis time, before the Spring is
    past.

The Summer's glories shine
  From all her garden ground,
  With lilies prankt around
And roses fine;
But the pink blooms or white upon the bursting trees,
Primrose and violet sweet, what charm has June like
    these?

MAY-TIDE.

    This is the time of song;
      From many a joyous throat,
    Mute all the dull year long,
      Soars love's clear note.
Summer is dumb, and faint with dust and heat:
This is the mirthful time, when every sound is sweet.

    Fair day of larger light,
      Life's own appointed hour,
    Young souls bud forth in white;
      The world's a-flower.
Thrill youthful heart, soar upward limpid voice!
Blossoming time is here: rejoice! rejoice! rejoice!
                      Lewis Morris.

THE CROCUS.

Like lilac flame its color glows,
   Tender and yet so clearly bright,
That all for miles and miles about
The splendid meadow shineth out,
And far-off village children shout
   To see the welcome sight.
Mary Howitt.

From EASTER MESSENGERS.

The Jonquil and the Crocus
   Have climbed up from the mould,
And scattered broadcast over earth
   Their amethyst and gold.
Like patient little miners
   Delving in frozen ground,
They gladden all the shivering land
   With treasures they have found.
Lucy Larcom.

TO A CROCUS.

BLOOMING BENEATH A WALLFLOWER.

WELCOME, wild harbinger of Spring!
   To this small nook of earth;
Feeling and fancy fondly cling
   Round thoughts which owe their birth
To thee, and to the humble spot
Where chance has fixed thy lowly lot.

To thee,—for thy rich golden bloom,
   Like heaven's fair bow on high,
Portends, amid surrounding gloom,
   That brighter hours draw nigh,
When blossoms of more varied dyes
Shall ope their tints to warmer skies.

TO A CROCUS.

Yet not the lily, nor the rose,
   Though fairer far they be,
Can more delightful thoughts disclose
   Than I derive from thee:
The eye their beauty may prefer;
The heart is thy interpreter!

Methinks in thy fair flower is seen,
   By those whose fancies roam,
An emblem of that leaf of green
   The faithful dove brought home,
When o'er the world of waters dark
Were driven the inmates of the ark.

That leaf betokened freedom nigh
   To mournful captives there;
Thy flower foretells a sunnier sky,
   And chides the dark despair,
By Winter's chilling influence flung
O'er spirits sunk, and nerves unstrung.

And sweetly has kind Nature's hand
   Assigned thy dwelling-place
Beneath a flower whose blooms expand
   With fond congenial grace,
On many a desolated pile,
Brightening decay with beauty's smile.

TO A CROCUS.

Thine is the flower of Hope, whose hue
   Is bright with coming joy;
The wallflower's that of Faith, too true
   For ruin to destroy;—
And where, O! where should Hope up-spring
But under Faith's protecting wing.
Bernard Barton.

PINK AND WHITE AZALEAS.

SONG IN PRAISE OF SPRING.

When the wind blows
  In the sweet rose-tree,
And the cow lows
  On the fragrant lea,
And the stream flows
  All bright and free,
    'Tis not for thee, 'tis not for me;
'Tis not for any *one* here, I trow.
  The gentle wind bloweth,
  The happy cow loweth,
  The merry stream floweth,
For all below!
  Oh, the Spring! the bountiful Spring!
  She shineth and smileth on everything!

Where comes the sheep?
  To the rich man's moor.
Where cometh sleep?
  To the bed that's poor.

SONG IN PRAISE OF SPRING.

    Peasants must weep,
       And kings endure;
       That is a fate that none can cure!
Yet Spring does all she can, I trow:
       She brings the bright hours,
       She weaves the sweet flowers,
       She dresseth her bowers,
For all below!
   Oh, the Spring! the bountiful Spring!
   She shineth and smileth on everything!
                       Barry Cornwall.

FLOWERS.

There are flowers round about me
  As I sit beneath the lime;
Sweet lowly things are breathing
  The breath of olden time.

They look so kindly upward,
  I greet them as my friends;
And my mind to each small blossom
  Such holy beauty lends,

That, as if to living creatures,
  Where'er my glance may fall,
On the blue-bells or the daisies,
  I say, "God bless you all!"

Go forth, my little daughter,
  The mid-day heat is o'er,

FLOWERS.

Go forth among the flowers,
   And gather thee a store.

The little fairy Speedwell,
   With its many eyes of blue,
How well I can remember
   Green lanes wherein it grew.

The Daisies, see how gayly
   Like little stars they shine,
The darlings of thy childhood,
   As once they were of mine.

The Blue-bell—when I see it,
   My thoughts fly back once more,
To a pine-wood, whose recesses
   With its bloom were purpled o'er.

Go forth, dear child, and pluck them,
   And bring thy spoils to me;
Thou lov'st the gay, bright colors,
   Though thou seest not what I see!

To me they bring remembrance
   Of many long past Springs;

They are types to me and shadows
   Of yet more lovely things.

They have sprung in joyous beauty
   From the drear and Wintry earth,
When all was dead and dreary,
   They have brought their new-born mirth.

Their stems are weak and fragile,
   To the faintest wind they bend,
Yet their coming is a token
   That death is not our end.

Not more of love than wisdom
   Was theirs, who round the tomb
First brought, in faith far-seeing,
   Gay flowers to bud and bloom.

On every leaf is written
   A sweet consoling thought;
The hope of life upspringing
   From death, by them is brought.

My child, my happy darling,
   Go pluck me many a one,

FLOWERS.

Though thou'rt the gayest flower
　That smiles beneath the sun!

Go forth, thou blessed being,
　And bring thy sweet spoils here,
Though I need no other token
　Of Heaven, when thou art near!

I need no other token
　Than thy fair and happy face,
Through which on me are beaming
　God's mercy and God's grace.
Anonymous.

WHITE LILIES.
(111)

Each with some holy message
    Unto these hearts of ours,
Through the dark doorways of the year
    Look forth the Easter flowers.
Soft floats their angel-chorus
    From fields and gardens dim;—
"The Lord is risen! Soul, rise and bloom
    A messenger for Him!"

                    Lucy Larcom.

A DIALOGUE FROM SOUL GARDENING.

"Thou bearest flowers within Thy hand,
  Thou wearest on Thy breast
A flower; now tell me which of these
  Thy flowers, Thou lovest best;
Which wilt Thou gather to Thy heart
  Beloved above the rest?"

"Should I not love my flowers,
  My flowers that bloom and pine,
Unseen, unsought, unwatched for hours
  By any eye but Mine?
Should I not love my flowers?
  I love my Lilies tall,
My Marigold with constant eyes,
Each flower that blows, each flower that dies
  To Me, I love them all.
I gather to a Heavenly bower
  My Roses fair and sweet;
I hide within my breast the flower
  That grows beside my feet."

Dora Greenwell.

LILIES.

We are Lilies fair,
   The flower of virgin light;
Nature held us forth and said,
   "Lo! my thoughts of white."

Ever since then, angels
   Hold us in their hands;
You may see them where they take
   In pictures their sweet stands.

Like the garden's angels
   Also do we seem,
And not the less for being crown'd
   With a golden dream.

Could you see around us
   The enamored air,
You would see it pale with bliss
   To hold a thing so fair.
Leigh Hunt.

THE LILY.

The stream with languid murmur creeps
  In Lumin's flowery vale:
Beneath the dew the Lily weeps,
  Slow waving to the gale.

"Cease, restless gale!" it seems to say,
  "Nor wake me with thy sighing!
The hours of my vernal day
  On rapid wings are flying."

"To-morrow shall the traveler come
  Who late beheld me blooming;
His searching eye shall vainly roam
  The dreary vale of Lumin."

Coleridge.

I SEND THE LILIES GIVEN TO ME.

I SEND the lilies given to me,
   Though, long before thy hand they touch,
I know that they must withered be;
   But yet reject them not as such:
For I have cherished them as dear,
   Because they yet may meet thine eye,
And guide thy soul to mine even here,
   When thou behold'st them drooping nigh,
And know'st them gathered by the Rhine,
And offered from my heart to thine!

The river nobly foams and flows,
   The charm of this enchanted ground,
And all its thousand turns disclose
   Some fresher beauty varying round;
The haughtiest breast its wish might bound,
   Through life to dwell delighted here;
Nor could on earth a spot be found
   To nature and to me so dear.
Could thy dear eyes, in following mine,
Still sweeten more these banks of Rhine.

Byron.

PALE YELLOW ROSES.
(121)

STANZA.

As when a lady, walking Flora's bower,
Picks here a pink, and there a gilly flower,
Now plucks a violet from her purple bed,
And then a primrose—the year's maidenhead;
There nips the brier, here the lover's pansy,
Shifting her dainty pleasures with her fancy—
This on her arms, and that she lists to wear
Upon the borders of her curious hair;
At length a rosebud—passing all the rest—
She plucks, and bosoms on her breast.

Quarles.

O HAVE you seen, bath'd in the morning dew?
　The budding rose its infant bloom display?
When first the virgin tints unfold to view,
　It shrinks and scarcely trusts the blaze of day.
　　　　　　　　　　　　　L. E. Landon.

THE rose said, "Let but this long rain be past,
　And I shall feel my sweetness in the sun,
And pour its fullness into life at last."
　But when the rain was done,
But when dawn sparkled through unclouded air,
　　　　She was not there.
　　　　　　　　　　Augusta Webster.

THE FALLING ROSE.

Pass, falling rose!
Not now the glory of the Spring is round thee;
  Not now the air of Summer round thee blows;
Pallid and chill the Autumn's mists have found thee!
    Pass, falling rose!

Pass, falling rose!
Where are the songs that wooed thy glad unfolding?
  Only the South the wood-dove's soft wail knows;
Far southern eaves the swallow's nest are holding;
    Pass, falling rose!

Pass, falling rose!
Linger thy blooms to birth thy glory wooing?
  Linger the hues that lured thee to disclose?
Long, long, their leaves the dark earth have been
    strewing;
    Pass, falling rose!
William Cox Bennett.

Woo on, with odor wooing me,
  Faint rose, with fading core;
For God's rose-thought that blooms in thee
  Will bloom for evermore.
                    George MacDonald.

Live all thy sweet life through,
  Sweet rose, dew-sprent,
Drop down thine evening dew
To gather it anew
When day is bright:
  I fancy thou wast meant
Chiefly to give delight.
                    Christina G. Rossetti.

www.ingramcontent.com/pod-product-compliance
Lightning Source LLC
Chambersburg PA
CBHW020156170426
43199CB00010B/1062